THIS BOOK BELONGS TO

Bernice

Burner

Elements of Evil

Elements of Evil

by Brooke Arnold

ThunderStone Books
Las Vegas, Nevada

Edited, designed, typeset, and project managed by Robert and Rachel Noorda at ThunderStone Books. Printed and bound by IngramSpark.

978-1-63411-005-1 (ISBN 13)

For the kids at home. You laughed in all the right places.

Measurement Conversions

Length
1 foot = 12 inches
1 yard = 3 feet
1 mile = 5,280 feet
1 inch = 2.54 centimeters
1 foot = 30.48 centimeters
1 meter = 3.2808 feet
1 kilometer = 1,000 meters
1 kilometer = 0.621 miles
1 mile = 1.6103 kilometers

Volume
1 pint = 2 cups
1 quart = 2 pints
1 gallon = 4 quarts

Weight
1 pound = 16 ounces
1 ton = 2,000 pounds
1 ounce = 28.35 grams
1 kilogram = 1,000 grams
1 pound = 0.4536 kilograms
1 kilogram = 2.2046 pounds

Temperature
Celsius (C) = 5/9 * (F-32)
Fahrenheit (F) = (9/5 * C) + 32

Newton's Laws of Motion

1. An object at rest tends to remain at rest, and an object in motion tends to remain in motion.

2. Force is equal to mass times acceleration.

3. Every action has an equal and opposite reaction.

Periodic Table

1 H								
3 Li	4 Be							
11 Na	12 Mg							
19 K	20 Ca	21 Sc	22 Ti	23 V	24 Cr	25 Mn	26 Fe	27 Co
37 Rb	38 Sr	39 Y	40 Zr	41 Nb	42 Mo	43 Tc	44 Ru	45 Rh
55 Cs	56 Ba	*	72 Hf	73 Ta	74 W	75 Re	76 Os	77 Ir
87 Fr	88 Ra	**	104 Rf	105 Db	106 Sg	107 Bh	108 Hs	109 Mt

*Lanthanides	57 La	58 Ce	59 Pr	60 Nd	61 Pm	62 Sm
**Actinides	89 Ac	90 Th	91 Pa	92 U	93 Np	94 Po

# of Elements

								2 He
			5 B	6 C	7 N	8 O	9 F	10 Ne
			13 Al	14 Si	15 P	16 S	17 Cl	18 Ar
28 Ni	29 Cu	30 Zn	31 Ga	32 Ge	33 As	34 Se	35 Br	36 Kr
46 Pd	47 Ag	48 Cd	49 In	50 Sn	51 Sb	52 Te	53 I	54 Xe
78 Pt	79 Au	80 Hg	81 Tl	82 Pb	83 Bi	84 Po	85 At	86 Rn
110 Ds	111 Rg	112 Cn	113 Uut	114 Fl	115 Uup	116 Lv	117 Uus	118 Uuo

63 Eu	64 Gd	65 Tb	66 Dy	67 Ho	68 Er	69 Tm	70 Yb	71 Lu
95 Am	96 Cm	97 Bk	98 Cf	99 Es	100 Fm	101 Ma	102 No	103 Lr

Equipped with his five senses, man explores the universe around him and calls the adventure Science.

—Edwin Hubble

To raise new questions, new possibilities, to regard old problems from a new angle, requires creative imagination and marks real advance in science.

—Albert Einstein

Science is a way of thinking much more than it is a body of knowledge.

—Carl Sagan

From: Bernice Burner
Sent: Today at 2:27 PM
To: Alec Burner
Subject: Thanks for the gifts!

Dear Uncle Alec,

2,$(3-53)^2$,8! 66-39,8 39,8,92 3,19,4-5 25-7,39 10,74
16,19,(3),3-53? 89-6 34,24,(52) 27,66-39,4-5 Be
92,16 He 92,34!!

I've been working on this all week. Isn't it awesome?

Thanks for the notebook. It's so shiny and new that I'm afraid to write in it! Bunsen loves his exercise wheel. He's gotten so big since you left him with me, you'd hardly recognize him. I love him, he's the best friend ever.

Mom says you're heading to South Africa next. Don't get eaten by wild animals! Then I'd have no one to discuss my ideas with.

Love, Bernice

From: Alec Burner
Sent: Today at 4:30 PM
To: Bernice Burner
Subject: RE: Thanks for the gifts!

Dear Bernice,

6,8^2,3-53 27,66-39,4-5! 101<-(10), 4-5,3-53,$(4-5)^2$,23 74,8,92,3-53,66-39 4 15,44<-8,105-5.

I'm glad you like the lab book! Don't be afraid to fill it up. All the smartest people write everything down. And I'm glad Bunsen likes his gift. Good to hear you two have bonded — I've always found hedgehogs make the best pets.

As per your request, I hereby promise not to get eaten by any ravenous, raging predators in South Africa.

Take care! Give your parents and sister my love.

Your favorite uncle,
Alec

~~Dear Diary,~~ ~~To Whom It May Concern~~, Dear Bunsen,
(Writing to a hedgehog trumps writing to nobody.)

First science club meeting of the semester! The anticipation was almost tangible as we waited for Mr. Science to show. He'd spent class dropping hints about some big news, and we were dying for him to tell us: he always makes such a production of making announcements.

Sure enough, he strolled in wearing his "I know something you don't know" smirk, dragging an enormous old gray chalkboard behind him. He spun it around with a flourish, grinning widely, his hair standing on end like it was electrified. The other side was covered in ribbons, blue and

4,16,22-53 52,89,1,68.
4,16,22-53 (10),102,92,7,58,109,←(10)

16,8 90,68⇐85,53,6,(57) 2 16,1,8,(71),66-39 52,89,1 66-39,88,(95)!

red and yellow, all surrounding bold white chalky words: COUNTY SCIENCE FAIR. Yes!

"I know what you're all thinking," he declared. "I need one of those ribbons. My future happiness, my very life, depends on possessing something of that very hue, sheen, and texture. Well, my scientific minions, it will challenge you. You may lose sleep, you may lose your hair, you may lose your lunch. But in the end, you may win something even greater. Eternal fame and glory, the prestige of being one of the few proud souls who have made their mark on the scientific community: a science fair winner."

What a ham. I love it. "Science" of course isn't his real name, but he rolled into class the first day wearing his flashiest tie and told us his last name is "an unpronounceable Italian mess" and that we would take a vote on what to call him this year. Awesome.

Back to today though — "I have in my hand your ticket to success: a signup sheet. Please, no stampedes. I will place this on my classroom door, for your perusal and delight after our meeting. We will discuss the fair details at length on a future date. Now, who's ready to make some carbon dioxide bubbles?"

Best meeting ever! Too bad club meeting's not every day. Usually there's nothing to do while I wait for basketball practice to end, so I sit like a lump. A boring lump. Joining science club was brilliant of me. No more boring waiting.

Today I had so much to do that I had to scamper to catch the late bus home! When I burst through the doors I saw Edith stalling, chatting cheerfully with the driver and glancing over her shoulder for me. As I stumbled up the steps the smile slipped off the driver's face and I got a stern "this bus is on a tight schedule" look. Figures. It was only at dinner when Mom asked about school that I realized: in my hurry, I forgot to sign up for the science fair! I have to remedy that as soon as possible, before it's too late!

This is the year. I can feel it. My time, my place! Edith isn't participating, so there's no one to steal my glory. Her science class doesn't require it, and anyway, she's too busy worrying about her English class. Apparently it's a doozy. I don't know why she's worried. She's going to ace it, just like everything else. But it's great news for me, because no competition! Just gotta sign up tomorrow. And choose a stellar plan. With all the brilliant ideas bouncing around my head, this fair's in the bag, and I'll finally have an award to put on my wall too.

Lab Notes: Science Fair Ideas

- Homemade crystals out of different things
- Best paper airplane designs and materials
- Training hedgehogs
- Circuit board Morse code
- Telegraph
- Resonators
- Botany — speed up your plant's clock
- Waterproof fabric

Dear Bunsen,

I spent most of lunch hour brainstorming which ideas I could use this year. I was originally tempted to submit a variation on last year, but the pain is too fresh. Something new, for sure. A fresh start is always better, wouldn't you agree?

I've been dying to tell Mom and Dad about how I'm going to blow everyone out of the water at the fair this year. But I've been waiting until I could come up with some solid topics, since naturally they'd want to know how I'm going to take the fair by storm, and that requires a plan of action. And of course, until they announced the fair, I couldn't be sure what I'll be working with. But since the announcement and coming up with plans at lunch, dinner today was the perfect opportunity. Edith finished making us all laugh with a comical story about saving her Chemistry class when the teacher's flaming-balloon experiment blew up, and then I felt it, the opportune moment.

"Speaking of chemistry, guess what! The science fair's coming up again," I told them.

And you know what happened?

"Oh, the science fair? That's so fun, Bernice. Remember last year, Edith's win? I remember, there was that spill that ruined your disappearing paint and put you out of the running. It was a real shame. I'm glad you're signing up again. Are you going to enter this year too, Edith?"

No questions for me. None. Just a sour reminder not to set up next to an experiment featuring bleach. Never again.

Anyway, Edith replied that no, she isn't entering this year, and the conversation continued like I wasn't even there. That's when it hit me. No one around here appreciates me. Three guesses why.

Edith.

25-7,39 14,16,22-53,68...

90 15,68,26,6,22-53 80-1,77,3-53.

60 96-6,4-5.

Edith does everything, and does it better than anyone else. Basketball, yearbook committee, chess club, student council, orchestra AND band, and of course science fairs.

Meanwhile, I suffer the repercussions. I'm made of hand-me-downs: clothes, classes, hobbies. By the time I get them, it's old news. No matter that I'm good at science, or have a great personality. Edith wore them first and I get the leftovers. The only thing I have of my very own is you.

But you know what? Not a problem. I've figured out the perfect way to stand out. A secret way. Something Edith's never tried. It plays to all my strengths: inventing super-useful gadgets, planning, dressing up. It's perfect!

Get ready for the new me … a supervillain!!!

90 10,74 25-7,4-5!
74,8,7'22-53 53 3-53,8^2,19 9,(11),73,16,22,6??

Dear Bunsen,

I'm liking this supervillain scheme more and more. It's a surefire way to stand out. And then I'll win the science fair, and that will be the icing on the cake! People are bound to pay attention when I swoop in and steal the show at the fair. No matter that they're distracted now ... in a few short weeks they'll see. No more standing in the shadow of Edith's success. Between the science fair and villainy, I'll be unstoppable. Bernice for the win!

Every good scientist does their research, so I'll start with some background reading, determine what others have to say about villainy and proceed from there. Read up on things before diving in. I've never tried anything like this before, and I'm not quite sure what comes first! So I've checked out a couple of library books to see what I can find. That should get me going.

Dear Bunsen,

What do you get when you cross a bubbly English teacher with the end of a unit on enduring friendship and loyalty? A mini essay, of course! Mrs. Reid had us do one of her ten-minute "descriptive exercises." This time we had to write about someone we love or admire. Her plan is for us to learn how to write more effectively by writing frequently and concisely, to "show, not tell" why this person is important to us. This one was actually fun — I wrote about you! I saved it to show you:

"The best days are the ones I spend playing with Bunsen.

He likes to watch me while I work. When I take him out of his cage he'll trot right over to where I'm working and cuddle up to my hands, listening intently while I tell him all the latest news. He's very attentive, and so interested in what I do. So interested, in fact, that after a few close calls with the Bunsen burner, I learned to keep everything remotely dangerous on a shelf out of reach. Too curious, but so clever! When I pack everything up, he trots right back to his cage to get ready for bed. It's our routine. Pack up, back in the cage, eat some mealworms, turn on the heater, turn out the lights.

Occasionally I get too busy to give him all the attention he deserves, which throws off our routine. When that happens, he misses me, I can tell. To him, I'm the most important person in the whole world."

Wasn't that great? Oh! And she hinted that we're starting a new unit soon which should "inspire and stir us." She seems to believe that literature should move us to greatness. Sometimes it's a bit over the top, but she really knows her stuff. I'll keep you posted on what she has in store for us.

Dear Bunsen,

Today decided to be one of those horrible no-good days. The kind that slowly drags you down into an ominous dark pit and leaves you there to rot. To be fair, the morning started okay, nothing exciting, just overcast and dreary-looking. But lunch. Everything fell apart at lunch. I stayed late in Mr. Science's class as usual to help him feed his fish. When I left his room I saw how dark the sky had gotten, and as I walked outside the rain hit. Buckets and buckets of ceaseless rain, pouring from the sky. And I had to get all the way across the school to the cafeteria. Running between buildings kept me mostly dry. The floors are what did me in. I stepped inside and had just enough time to see everyone's heads turn toward me, and to glimpse Mr. Murphy busy janitoring halfway across the room, mopping up mud and slime. Then the world flipped, and I found myself lying facedown in the muck. In front of everyone. I was pretty stunned, so I just sat there trying to catch my breath, but a second later I felt hands grasp my elbows and lift me to my feet, and when my eyes focused I saw Mr. Murphy pulling a towel out and handing it to me. No one else even moved. Plus the whole school saw it, and I had to suffer through the rest of school and the bus ride home in clammy, wet, mud-clothes, with everyone avoiding looking at me. Then when I got home Mom saw me, and didn't react at all, just sent me upstairs to change. I forgot to give Mr. Murphy back his towel, too, so I have to go return it.

Then, to top it all off, Edith dropped a bomb at dinner. That English class she's been stressed about? The requirements changed for the mid-semester paper, which apparently freed up her whole life. And of course rather than basking in free time, she's decided to fill it. Competing in my science fair. The nerve! And I forgot to run my ideas by Mr. Science again. If I don't hurry, she'll get ahead!

As I was thinking about it tonight I came to an important conclusion. Days like today are yet another reason to pursue supervillainy. Everyone will admire me so much they won't dream of competing with me or care about things like muddy clothes.

Dear Bunsen,

Obviously after yesterday's setbacks, the mud debacle and Edith's declaration, I was incapable of progressing on my science fair project or anything else, but today was smooth as silk, and I'm back on track. No more meltdowns, just pure focus. Like a tiger. It's all coming together.

In other interesting news, today Mrs. Reid informed us that in the next unit we'll be discussing heroes and villains in literature and media. Seems really neat. We'll be comparing movies, books, plays, and art and discussing what roles each character plays and why. Movies for homework? Done. The whole class was really into the idea. Of course, she chose that moment to shatter our illusion, informing us that we'll be finishing the unit with a research paper on heroes. No free ride after all. Still, it's a cool idea. We have to include at least one interview and one news event as well. The thing is, it'll probably eat into my science fair time, but who knows? Maybe I can tie them together somehow. Heroes and villains: an in-depth study. The science of heroics. The science of villainy!

I also finally talked to Mr. Science about my science fair ideas when I was feeding the fish at lunch today. I'd planned to yesterday, but you know how that went. I almost forgot again today. He was explaining how fireflies work — there's this chemical reaction that takes place and produces a bioluminescent enzyme which gives off cold light, but the coolest part is that the firefly controls the reaction by controlling oxygen flow! Wow! You know how I get when people explain things. But luckily he brought up the fair. He said he's planning a "board review" of our planned course of action for next week's club meeting, so I can run my ideas by him then. I do worry a little that waiting until next week will delay my progress too much. Maybe what I'll do is give Uncle Alec a call to ask his advice. He studied science in college, you know, and he taught me everything I know. He'll give me some great pointers!

27,7,22-53,(103)←8
18←53 – 3,(80),22-53
28,(80),22-53

The one thing I didn't do yet is return Mr. Murphy's towel. I was thinking it over, and I kind of want to thank him for giving me a hand. And a towel. He was the only person who came to my rescue. The rest of the room sat by dumbly while he ran to my aid. Such a thoughtful gesture deserves a little appreciation. Because of that I decided to wait to return his towel until I can think of a way to thank him.

In other news, I've gotten a start on my secret villainous plan. That's coming along swimmingly. I learned a ton from my reading!

Villain Notes: Book Research Results

After some reading, it is clear the first step to becoming a supervillain is finding a suitable nemesis. I found this in *Supervillains: All You Need to Know*,

"Every villain has one. It is crucial to select a nemesis hero carefully, and not just grab the first one that appears. A well-chosen hero shows off the villain's talents, challenging him to rise above mere excellence. They have to complement each other, comparable in both intelligence and understanding. Otherwise villainy becomes all work and no play. How can you have a witty exchange if the hero has no sense of humor or irony? What if he is already paired with a supervillain? What if his favorite color clashes with yours? What if he's a jerk? These are the questions to ask. Never make a serious commitment like that without some background information."

It's a superb point — selecting a nemesis could be a tricky business. I've reflected on this as I've been reading, and I think the best place to start is to go straight to the source: track down some heroes and start asking questions. To this end, I have decided to prepare an ad which I will place online to gather in potential candidates. It requests any and all heroes in the immediate area to participate in an interview for a project to showcase their talents. I'll generate an initial pool of nemesis candidates by conducting interviews. That'll be a good place to start. I'm not sure yet how I can meet them though. I'll need a way to safely meet with a bunch of strangers. Preferably without having to sneak around to do it. Some public place, no dark alleys. I don't want to die an early death!

Dear Bunsen,

Uncle Alec liked my ideas for the fair. He says I just need to come up with some questions to go along with the general topics, and that'll give me a good foundation to create the prize-winning project. He did say that training hedgehogs and the botany one could be pretty time-intensive, so I think I'll set those aside. A good start!

Dear Bunsen,

It occurs to me I need interviews for Mrs. Reid's paper. It's a perfect opportunity! Interviewing heroes to find a nemesis, under the pretense of interviewing them for school? What could be more devious than that? So I dropped some hints and Mom promised to help me reserve a room in the library to do a group interview on Saturday with whoever shows up. I submitted the ad today, and it looks pretty good, if I say so myself. Solving two problems with one round of interviews, score! Here's to research!

HEROES WANTED!

Is saving the day your specialty?
Help educate humanity by
participating in a personal
interview!
Come talk about your success,
talents, and background.
Saturday. 6PM @ the Library.
Share your story.

Dear Bunsen,

Uncle Alec stopped by tonight! He's back from Africa, but has to travel again right away and was on his way to drop my cousin Cooper off with his mom's parents. When Alec's travelling we all trade off on who keeps Cooper, since he's too young to go all over the globe every few weeks. Side note, when Uncle Alec's at home they live in this awesome house, with a darkroom for developing photos and a lab for experiments. Coolest place ever. Anyway, they only stayed long enough to give everyone a hug and say where he's going first (photographing the Great Barrier Reef!) before they had to run. But he promised to send me mail while he's journeying. Yay! I love when Uncle Alec travels.

Dear Bunsen,

Today was the day. Interview day. Mom brought me to the library and helped me check in with the librarian and find a room to set up in, then left me to my own devices. I put up a sign, "hero interviews here," and waited, ready for anything. Or so I thought.

The interviews were a disaster. I kept recording for as many people as showed up — I think there were seven or eight in the end. I didn't check. The first couple should give you a sense of the rest:

The first person was small and weedy, with an ego to make up for it. His first words to me? "Today is your lucky day; I should be an enormous help with your little research project." Longest two minutes and twenty-six seconds of my life.

The next guy was enormous. Bigger than the space shuttle. He didn't seem very bright, but was plenty cocky. "Villains are committing all sorts of nefarious crimes here: planting cats in trees, making it snow in May, playing the same song over and over on the radio. Evil is everywhere. And I will stop it."

It was too ridiculous: a bunch of arrogant wannabes. I kept hoping the next person would be better. Aren't heroes supposed to be more heroic than that? Especially grown-up heroes! Obviously I went wrong somewhere. Advertising brought a bunch of fakes. So where do I find the kind of hero Mrs. Reid always talks about? I'm going to have to change my tactics somehow, that's for sure. Maybe I'll ask for some advice tomorrow.

Dear Bunsen,

Managed to snag Mrs. Reid before she finished packing up after class, and explained what a flop my interview was. Such a debacle, and it stumped me. So I decided to go in and ask her advice on finding a real-life hero. I explained what happened at the library and how I wasn't sure what to do next, and she gave me a look.

"Those people have no idea what being a hero means. Wrapped up in themselves, absorbed by the idea of glory and spend their whole time looking for it. Real literary heroes are the opposite. They start off in the shadows and then rise to meet challenges that come their way. They don't go looking for it. It comes to them. You need to look for people who rise above challenges, against all odds. If you find the problems, you'll find the people."

Bunsen, I hadn't thought of that. I mean, not in so many words. But she's right. Real heroes aren't looking for it, they just see problems and step into the role. They're *molded.* Molded by the challenges they face, and a hero with certain talents for solving a problem will become a certain type of hero. They become the kind of person people look to in situations because of how they've handled similar ones.

You know what this means, don't you? I can't expect a hero to just fall into my lap. I'll have to mold one to suit. And I think I know how.

Villain Notes: The Game Plan

The likelihood of just *finding* the perfect hero is pretty slim. I am just a kid, so I have to set my sights close to home. And even if the perfect nemesis exists, I can't be expected to travel the globe to find him. Mrs. Reid's insight will be the key: I just need to find someone with potential, and mold him to fit. We'll be perfect. Probably someone at school would be best, where I can monitor his activity ... In the meantime, I realized I've been taking the wrong approach to this whole thing. How do villains start off? They attract attention, build themselves a reputation, and heroes flock to them!

I think this will have to be a three-phase plan.

Phase One: Build a reputation.
Phase Two: Find a hero.
Phase Three: Put the two together to make a big debut!

Dear Bunsen,

I finally returned Mr. Murphy's towel. Or tried to. He was off janitoring when I visited his office. I couldn't think of any good way to thank him, so I just wrote him a little note and planned to give it to him with the towel. Tracking him down was an adventure though. He has his own personal office, but it's tucked away in the darkest corner of the school where only mole people would ever find it. I finally had to ask directions from the secretary in the main office, and got lost twice on the way. And then he wasn't even there. But when no one answered my knock I tried the door, and it was unlocked. Not very secure, but convenient! I poked my head in and beheld a breathtaking sight: all kinds of useful knick-knacks, all neatly lined up along the walls. Every wrench and screwdriver imaginable, and a million boxes heaped with screws, nuts, bolts, washers, nails. It was like a mini garage: the cleanest, most beautiful garage I've ever seen. Since finding this place had been harder than wrestling a tiger I figured it was perfectly okay to go in and

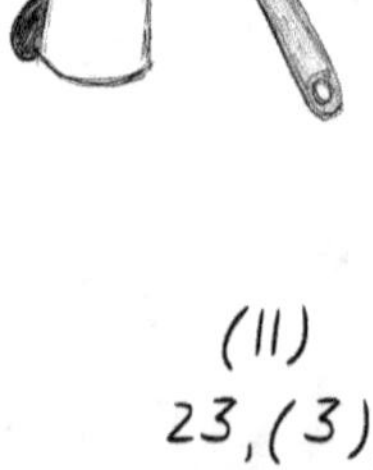

leave the note and towel behind. I set them on his desk where he couldn't miss them, and left in a hurry. I just have to say, wow. What a treasure of useful materials, and well-organized and tidy too! Someone with an office like that deserves some serious respect. I know where to go if I ever need … anything!

(11) 33,15,77,49,80-1

23,(3),57,49'16 66-39,75,95

Dear Bunsen,

A no-school-day, hurray! It was raining last night, but as some point it became freezing rain, which is quite the phenomenon.

Here's how it works (I did some reading on meteorology this morning):

There's a thin layer of cold air near the ground surface, sometimes caused by inversions. That's where it's been cold, like happens after a snowstorm, but then the sun comes out and warmer air traps the cold air in a pocket so it can't escape. The rain stays liquid because the cold layer under the warm layer is so thin, then supercools on contact with the cold ground and forms a layer of ice.

Freezing rain is awesome and dangerous. It makes black ice and can cause accidents and power outages as the ice builds on trees and power lines. Naturally, the school decided to play it safe and closed for the day. Staying home meant I could hole up in my room and really focus on making some headway on the science fair. Plus tidy a little. Can't change the world in a pigsty.

Sitting at home all day also gave me lots of time to think and plan evil plans. I have a great idea for getting started.

Villain Notes: Phase One

Time to start stirring the pot. Nothing too dramatic to begin. I mostly need to test the waters, find out what I can get away with. Then I can focus on really ramping up my efforts. Setting up a trademark. Sending a message. I can watch who puts an effort into tracking me down and preventing incidents, too. That could give me a sense of some heroes. Just have to perfect my plan for the first event.

Lab Notes: Icy Ideas

What can you do with ice? Supercooling is neat. Water on a freezing surface too — like black ice, or an iced-shut door like happens in ice storms.

Dear Bunsen,

I was looking for some inspiration, so I emailed Uncle Alec asking if he'd ever used his science experience to prank people. There's never a guarantee he'll respond when he's on travel, but I lucked out! It gave me just enough to polish off the first part of phase one.

From:	Alec Burner
Sent:	Yesterday at 11:13 PM
To:	Bernice Burner
Subject:	RE: Science Pranks

Bernice -

Pranks, huh? Yeah, when we were young I was always looking for ways to annoy your dad. He played soccer, so I would tamper with his water and drinks. Adding chemicals to them, things like that. His face was always priceless.

69<-53,4-5 He 44,7. 2<-[8,15] (22) 2,3-53,15,16!

Love ya!
Alec

Dear Bunsen,

Have I ever told you about Mr. Morales's fetish? I heard once that he was a track star in college or something, and he had to drink gallons of liquid every day. Well he's *obviously* no track star anymore, but I have never seen him without a sports drink either in his hand or half-empty on his desk. He's got an ancient fridge in his room, full of drinks. He also wants everyone to share his love of hydration, so he lets kids store their own drinks in the fridge as well, as long as they're clearly marked with a name and date. It's a good setup, if a little bizarre. I've used it a few times this year, and I know Edith and her friends love it. Sports people. Psh.

Villain Notes: My First Official Act

The key is to start small. The reason? To find out how easy it is to prepare an evil plan and put it in action. I want to see how much it will take to get people to take note of me. This first act isn't remotely dangerous, hardly villainous at all. Planting some specially prepared drinks in a certain teacher's fridge. I've already tested it — it's straightforward and replicable; time is the most important factor. The supplies were easy because we keep a healthy stock of sports drinks on hand for Edith. Assuming the results are positive, I can proceed to more evil deeds. I have to move quickly, to ensure people pay attention and make the connection. No more than a week. Results to follow.

Villain Notes: Phase One, Shaking Things Up

The Goal: Test the waters and confuse people by booby-trapping their drinks. Two different tricks should be good enough. If one fails, the other should cover it.

Trick #1: Supercool a drink past freezing to create a slush when shaken.
Problem: Maintaining temperature.
Solution: Cool the fridge to help it stay cold longer.

Trick #2: Altoids plus artificial sweetener makes the color disappear.
Problem: Requires adding things ahead of time.
Solution: Remove with the promise to restock.

Dear Bunsen,

I decided to stock up on lab supplies tonight. So after rinsing the last dish after dinner, I grabbed a bag and walked into the pantry. I started emptying shelves into the bag.

"Hey Bernie, Mom wants me to — Bernice?"

"In here! Just getting some stuff for my lab."

I stood, turning. Bang! The edge of the door is harder than you might think. Enough to rattle your brain if you hit it with your face. Stars bounced in front of my eyes as I dropped the bag and grabbed onto the doorframe. Ow…

"Bern, you okay?" A shadow stood in front of me. I shook my head as my vision cleared, turning the shadow into Edith, standing uncertainly with a sports drink in one hand, my bag in the other.

"Fine, just doing some physics. With my face. Testing Newton's laws. They're still in effect, no need to worry."

"Um, ow. Do you need some ice or anything?"

My head pulsed. "No, I'm fine. Fit as a panther."

"If you're sure …" She frowned, and held out the bag. I took it and held my hand out for the drink. "Oh, right. What experiment are you using these for?"

"That one's actually not for the lab, I'm just grabbing a drink to stick in Mr. Morales's fridge for tomorrow."

She nodded. "Oh! Before I forget, I came down to tell you, Mom's making a grocery list and wants to know if you need anything, for Bunsen or school."

"Umm …. I could use some more mealworms. And some bedding, it's time to clean out the cage."

"You know, Bern, hedgehogs are cool, but the whole mealworm thing … uggghhh." She shuddered and shook her head.

I let it pass this time. It's not like she has to eat them.

Note to self: buy a helmet.

49,6,71,66-39,4-5 2,3-53,25-7,(52) 49
23,(3),57,49 8,113-92,9,(22)

Dear Bunsen,

Science fair followup! I haven't made much progress on deciding on a final idea, what with one thing and another, so I was really glad that Mr. Science was going to do another board review. He of course waltzed in with his usual flair.

"Good afternoon, my little organic masses. I hope you've come prepared, as discussed, to amaze and enlighten me with your scientific prowess, for today is board review of your progress so far on your project plans. You have three minutes to write down your progress and submit it to me. Start … now!"

We all scrambled to pull out paper and pencils, and you'd be surprised how intense the next three minutes were, five people pen-tapping/chewing/scratching, while Mr. Science reclined in his chair, feet on his desk, grinning evilly. Bunsen, if there was ever a living embodiment of a mad scientist, he's it. He's an inspirational monster.

"Time! Rise and report!" He swept up our papers and skimmed each, then taped them to the board. "Brace yourselves, you may learn something today."

He proceeded to pick apart each person's report, acknowledging the merits, pointing out problems, and giving general advice. It was super enlightening! It's amazing how much he can see after just glancing through once.

He finished by telling us, "Overall, not bad. On a scale of baking-soda volcano to Nobel prize, you all fall solidly in the middle. A great start, and with all my illustrious advice, you may edge forward to become winners. See you next week."

When he mentioned the baking-soda volcano, I had this mental image of a volcano sitting among all the other projects like a little lost kindergartener on the big-kid playground … I wonder if anyone has ever submitted one, or at least a variation on it. I mean, they're flashy and messy enough to seem like a good starting point for someone, and so easy that someone's bound to have tried it. That would be hilarious, don't you think?

But anyway, I got some really good feedback. I have to mull it over a little, but I'll let you know what I decide.

60 90 74, 49, 7, 68 (14)...

Dear Bunsen,

It took an extra day because of the hiccup running into Edith in the pantry, but I just planted the drinks. Now it's just a matter of waiting.

Dear Bunsen,

Look, a postcard from Uncle Alec today! I wish I could tell him about my evil plot. I bet he'd have all sorts of good advice.

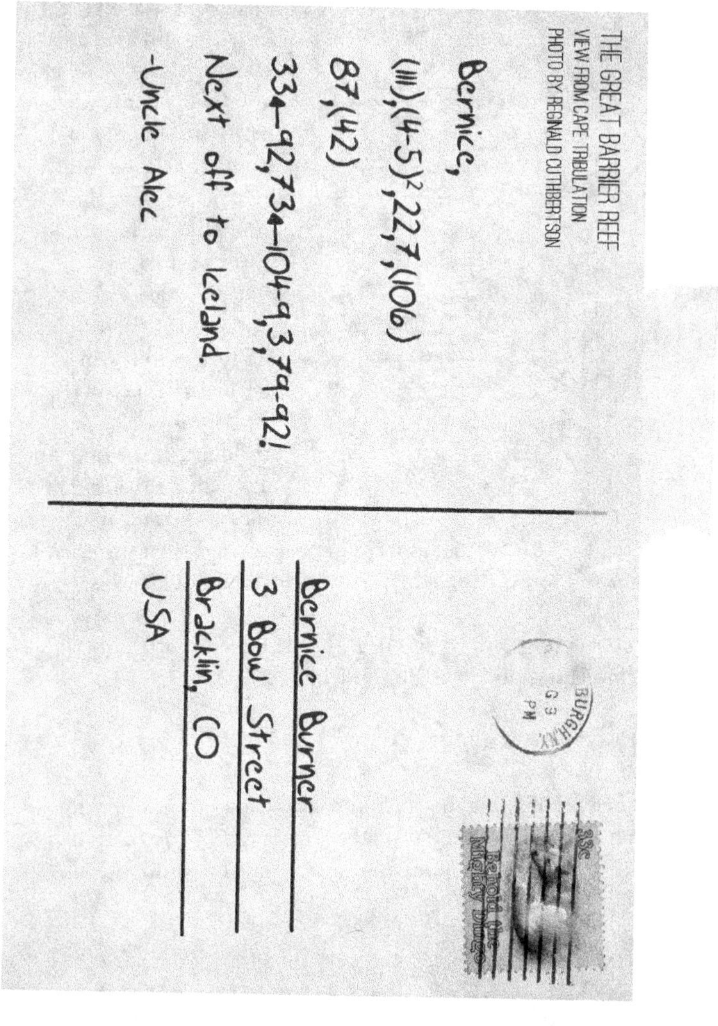
THE GREAT BARRIER REEF
VIEW FROM CAPE TRIBULATION
PHOTO BY REGINALD CUTHBERTSON
Bernice,
(111),(4-5)², 22, 7, (106)
87, (42)
334-92, 734-104-9, 3, 79-92!
Next off to Iceland.
-Uncle Alec
Bernice Burner
3 Bow Street
Bracklin, CO
USA

Dear Bunsen,

Edith was bursting with news when she met me at the bus stop today. I could tell. She tried to play it cool because I was reading (*History of the Periodic Table*, the one good thing that came out of that library visit), and perfect Edith would never interrupt, but she was practically vibrating with excitement, so I reluctantly set the book aside, and she immediately launched into the story.

Apparently there was an uproar today in her history class with Mr. Morales. As per his habit, he began class with a drink. He pulled one out of the fridge, took the cap off and took a swig, and then watched in confusion as his drink turned into slush before his eyes.

Disconcerting, but no big deal, he shrugged it off and started class, just a random mishap.

But that wasn't all.

At the end of class, he took another drink out of the fridge. He shook it up, maybe to make sure it didn't freeze up like the other. And the color disappeared! All he could do was stare in shock, and then the bell rang, and the class all left, and he was just staring flabbergasted at the bottle.

"How did they pull it off?? I wish I knew how those tricks worked. They were impressive. But the real question is, who did it? You were there yesterday. Did you see anything when you put your drink in?" Edith told me eagerly. "The slush drink could be a fluke, but the color changer? Someone planted that, for sure."

87,8,Z,(10) 49 16,1,8,6,19

Dear Bunsen,

I ran into Mr. Murphy again today. Walked by Mr. Morales's class on the way to lunch, and Mr. Murphy was there, looking at the fridge. He noticed me and waved me in.

"Hey kid. Thanks for the note a couple days ago. Such a nice thing to see after a long day. Just checking the settings on the fridge here. It was turned down to almost freezing … Quite the show here yesterday, huh? Neat trick with the supercooled slush, and then the color-changing juice too — that one was clever. Overheard the teachers chatting about it in the lounge this morning. They were speculating about it, thought some kid probably decided to show off to his friends. Took some planning to switch out the bottles, and maybe set up the supercooled drink too. Timing's got to be pretty good for that. The only thing I wonder is, what's the point? Was it just some kid showing off to his friends? But if so, why use the teacher's fridge? Why not just stick it in his friend's backpack and call it good? Makes you wonder, huh?"

Well, Bunsen, I didn't know what to say to that. He's right. It couldn't be someone showing off to his friends. I did my best to make a graceful exit, saying something eloquent like "Yeah, sure is a lot to consider."

That janitor's a smart one. I'll have to watch my step around him.

Then science club! Mr. Science practically danced into the room, he was so ready to tell us all about whatever he wanted to tell us about. When he skidded to a stop, he rubbed his hands gleefully. "Children! I came directly from the teachers' lounge, where I heard the most marvelous story! Of cold, chemistry, and calculation! So today, we are shaking things up a bit, and talking about ... drinks!" Shaking things up. Really! I think he sits around in his spare time coming up with puns, and the rest of his time he spends looking for places to use them.

He then proceeded to explain how to replicate the magical slush and disappearing color tricks that Mr. Morales must have told him about. How you can leave a drink in the freezer until it cools past its freezing point, so that when you shake it, ice crystals form instantaneously, and you get instant slush. And how the color in sports drinks comes from a chemical that, when mixed with the right ingredients, will vanish when shaken, only to reappear when you shake it again.

Villain Notes: Postmortem Assessment

The drink trick worked well, and gathered more notice than I'd expected. People noticed! They were talking about it afterwards, both students and teachers! Definitely a good sign. Now I need some way to begin making trouble, real villainy. The important thing now is scaling. Can I attract notice without getting caught as things get bigger? That's what the next step is all about: stepping it up a notch. That baking soda volcano idea? I can use that. As a tribute to Uncle Alec, who will be in Iceland any day now to photograph volcanoes. Plus, what perfect irony — stirring up trouble with a cliche trick.

Dear Bunsen,

I found out my class is assigned to lunch duty this week. Have I told you about lunch duty yet? Lunch duty is this thing our principal came up with a few years ago to teach us how to be "contributing members of society" by subjecting us to a few hours of community service a term. The community we're servicing is the school, and if we're being honest, it's glorified slave labor to reduce costs for hired help, but don't anyone tell the principal that. He's announced he's considering janitorial duty as well. That wouldn't be so bad, getting to work with Mr. Murphy, but janitorial? Yuck! Still, when you consider lunch duty ... a step up, for sure.

Kitchen duty works by selecting one English class a week to help run the kitchen for hot lunch. Six or seven kids go each day. They get out of their before-lunch class half an hour early and everyone reports to the cafeteria, where Ms. Chu is waiting. She's the epitome of the stereotypical scary lunch lady, from the hairnetted gray bun to the white orthopedic sneakers to the semi-transparent plastic food gloves on her hands, which are either crossed grumpily in front of her chest or planted in fists on her hips. I've never seen her crack a smile — her face would probably shatter like a porcelain vase if she did. She's always waiting with a blue bin full of yellowish-gray-stained aprons which were probably used by the dinosaurs for mud pies in their swamp homes, and you grab an apron and report to your assignment. You can be assigned cooking duty, dishes, cleanup, and serving. Dishes and cleanup are best. She hovers over the cooks and servers, presumably to make sure they don't make her look bad by poisoning anyone's food. Bunsen, it's very disconcerting to be minding your own business, stirring a pot of tomato

sauce, and then suddenly feel hot breath on your neck and a looming presence behind you. Makes you want to jump and accidentally fling the spoon full of steaming sauce into her face. But of course, that would get me stuffed into the spaghetti sauce with the mushrooms and parsley and then simmered until smooth and silky. So maybe not.

Cleanup though is no sweat. She's so busy terrorizing the other kids that you can pretty much do your own thing, as long as you don't get in her way.

Villain Notes: Round Two

Use baking soda and vinegar to create a kitchen surprise! Ketchup and baking soda will make an oozy mess. Ms. Chu is so uptight it should give her a fit!
Problem: time delay. Mixing it plain would end up with me caught red-handed!

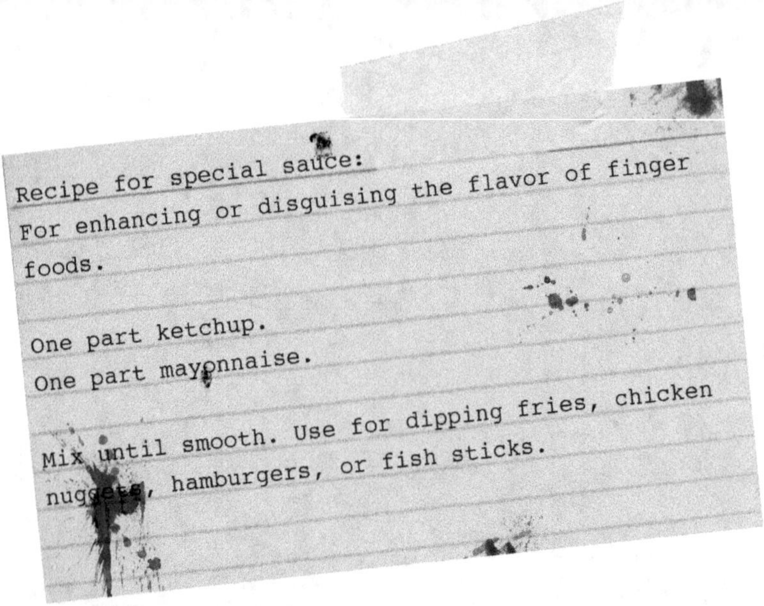
Recipe for special sauce:
For enhancing or disguising the flavor of finger foods.

One part ketchup.
One part mayonnaise.

Mix until smooth. Use for dipping fries, chicken nuggets, hamburgers, or fish sticks.

Solution: mix baking soda into mayonnaise. No reaction until mixed with ketchup!

Snagged this from Ms. Chu's secret recipe box, which is a locked box stored right by her keys, on a shelf by the kitchen door. Poor planning!

Villain Notes: Progress

Last day of lunch rotation. I waited until the last day because otherwise I would have had a hard time pulling it all off without getting caught: I can't be anywhere near the kitchen when this volcano erupts. Luckily I was on cleaning duty today, so Ms. Chu's attention was everywhere but on me as I casually swapped out a jar of mayonnaise for the soda-laced one. When another class is in charge next week, they'll be the first under scrutiny, and I will be under the radar. Airtight and foolproof.

Dear Bunsen,

Lunch today had quite a ruckus. A commotion. An uproar. Ms. Chu's howling could be heard all the way in the office — I know, because I was there carrying a message for Mr. Science, a note from a student getting out of class early. I had just handed it over when the most unearthly shrieks imaginable began. Think nails on a chalkboard plus microphone feedback. The secretary and I froze, then she leapt to her feet and we both ran, following the yelling to the lunchroom. When we arrived it was hard to tell what had happened, exactly. I edged around so I could see. No one noticed me. Everyone was too stunned, frozen in place while in the kitchen, this week's lunch crew were all cringing while Ms. Chu yelled at no one, red-faced, flinging her arms around, gesturing to a large bowl filled with bubbling pinkish ooze which I presume was supposed to be the special sauce for today's fries. When I could make out her words, which was naturally difficult, given her volume, speed, and pitch, you could hear "Sabotage! Trying to ruin me! Poisoning the lunch, putting me out of a job! I demand justice! You will pay for this, you, you …." I kept expecting her to run out of steam, but she just continued, an unstoppable banshee.

At this point Mr. Murphy arrived, inconspicuous in his gray shirt, a bucket full of cleaning supplies gently swinging by his side. He edged his way through the crowd to Ms. Chu and put his hand on her arm, narrowly missing being whacked in the face as she spun around to face him, her shrieks dying out as she saw who it was.

"Ma'am, if you could step aside, and help your crew out of the kitchen, I can take care of this mess." He calmly ushered everyone out. At this point, the lunch monitors returned to the present and began evacuating the cafeteria. I imagine they were worried about fumes or something. As I was swept outside with the crowd, I glanced back and saw Mr. Murphy considering the bucket with a thoughtful frown, hand on his radio. I wish I'd been there to see how he handled it, but no dice. All I heard was that it was taken care of, and there was no immediate danger to anyone.

Later:

Of course Edith dashed up to me at the bus stop the minute she got out of practice.

"Bern! I heard about lunch; I was reviewing a paper with Mrs. Reid and missed the whole thing! You were there, right? What happened?"

I gave her a short and sweet synopsis.

"Sounds like Mr. Murphy took care of it … I should talk to him, see what he thinks happened, see if there's any help he needs making sure things are safe. And I heard Ms. Chu is thinking of quitting now. She's taking the rest of the week off for sure. What a paranoid, cranky old lady. Looks like bag lunch for a couple days until we can sort it out."

Lovely, I didn't even think of that. Practically the whole school has hot lunch, and now they all have to change their ways. Because of me. What a great moment — this lunch volcano became a bigger deal than I'd anticipated. The whole school should take note of this.

I swung by Mr. Murphy's office after school to see what he had observed at lunch. I'm really beginning to appreciate basketball practice. So useful to have some extra time at school. Anyway, he looked surprised to see me, but was willing to chat when I explained why I was there. He didn't say much, just that he was able to clean it up, and that there were a couple things that made it seem like it was deliberate like Ms. Chu claimed, not an accident. But he didn't really elaborate. "I don't know anything for sure, so I'm hesitant to say too much right now."

Mopping up the Mess: Troubles at Lunch

CAFETERIA— Wednesday: chicken nugget day. As predictable as it is uneventful. But last week the humdrum routine of standing in line for lunch was shattered by screams. Ms. Chu, the stern and commanding kitchen supervisor, fled her domain in panic, screaming about sabotage. The lunchroom was quickly vacated with no further mishaps. The cause has not been revealed yet, though the janitor was on scene within moments, and the lunchroom was opened for use the following day. Ms. Chu has not since returned to work. Students have begun to wonder, how long will they have to bring their own lunches?

He's just like a mystery investigator. Cooler and cooler. He could very easily be the hero I'm looking for! Observant, quick to respond, smart ... definitely a hero. I think I can work with this!

Dear Bunsen,

Turns out Mr. Science agrees with Mr. Murphy about the recent activities. Science club today should have been titled "Detective Work Featuring the Scientific Method."

He rushed into the room after everyone was seated, and to all our surprise, Mr. Murphy followed on his heels, seating himself in the back of the room. "Today, we are going to talk about reactions. As you know, yesterday's lunch was was somewhat of a, shall we say, explosive experience due to a simple chemical reaction. Our guest today, Mr. Murphy, whom you should recognize as our janitor, called me in when he arrived on scene to determine the level of danger, and I recognized it instantly: a cleverly disguised baking soda volcano. Observe."

He pulled out a plate, a bottle of ketchup, and a container of baking soda from behind his desk. He squirted ketchup onto a plate, dusted it with baking soda, and waited. Nothing happened.

"Alas, no reaction. But watch."

He whisked out a popsicle stick and stirred. The ketchup slowly started bubbling, rising and oozing off the side of the plate.

"Behold, the power of vinegar and baking soda! May I remind you, one of the primary ingredients in ketchup is vinegar — that's where the kick comes from. Not dangerous, and not particularly spectacular by itself, but given the venue and Ms. Chu's particular tendency to, uh, react explosively to problems in her kitchen, it was certainly a phenomenon worthy of note. So that begs the question: what exactly happened here? Well, the scientific method should help us answer that!"

We spent the rest of the hour discussing the whole situation carefully, Mr. Science writing frantically on the board, Mr. Murphy sitting quietly in the back. Bunsen, he didn't say a word the whole time. I can't decide if he was there for moral support or to try and get some more insight on the ketchup volcano story, or what, and I had to run right after to get home with Edith, so I couldn't stick around to find out.

Notes from Science Club: Ketchup Volcano

Question:

What happened to cause a ketchup volcano in the kitchen?

Background:

No baking soda in sight, and last week there was the magic drinks incident

Observations:

- Ms. Chu monitors the cooks
- Chicken nugget day = ketchup + mayo mix for dipping
- No baking soda in sight
- All helpers claimed to know nothing
- New container of ketchup
- Clean kitchenware (carefully monitored)
- Old container of mayonnaise
- Middle of the week
- Reaction occurred while making dipping sauce
- Baking soda reacts with vinegar in ketchup

Hypothesis:

Someone created a "volcano time-bomb" with baking soda, ketchup, and mayo

Procedure:

- Create control mixture of ketchup, mayo
- Test 1 — control vs. mix of lunchroom ketchup, control mayo, and baking soda
- Test 2 — control vs. mix of control ketchup, lunchroom mayo, baking soda

Results:

- Control: no reaction.
- Test 1 — no reaction.
- Test 2 — bubbles!

Conclusion:

Planned and purposeful. Same person as the drink bandit?

"Being clever minions, I hardly need to tell you what I concluded. But I will. Someone stirred baking soda into the mayonnaise, to keep it isolated until the two were mixed. Probably a lot of baking soda, given the size of the reaction."

Villain Notes: Postmortem

Looks like making waves has started. Excellent. Just gotta stay under the radar — Mr. Science is smart, and Mr. Murphy's definitely on the lookout for the culprit. Plus Edith's sniffing around too now, and who knows who else. I'm glad to see people starting to take me seriously now. Progress.

Dear Bunsen,

Would Mr. Science make a good nemesis? I had Mr. Murphy in mind, but Mr. Science obviously has a talent for detective work. And he's excited to jump in and track down the source of all the hullabaloo. Young, energetic. Maybe too energetic though. He's really goofy too, and shouldn't a hero be kind of serious? I can't imagine people liking a hero who cracked jokes as he rescued a cat from a burning building, can you? No, he seems a bit flippant for a superhero. Still the best science teacher ever though. And he's certainly contributing to the effort. Works well with Mr. Murphy. In fact, maybe I'm not looking at a single nemesis after all. Maybe I need a team of heroes!

Thoughts: English Essay

- Qualities of a hero: likeable, selfless, successful?
- Who could be considered heroes?
- Mr. Science, Mr. Murphy, anyone else?
- Raising awareness of the dangers of pranking, and
- Putting together the puzzle of who is doing this and how to stop them
- Active cleanup and prevention
- Projecting what the villain might do next

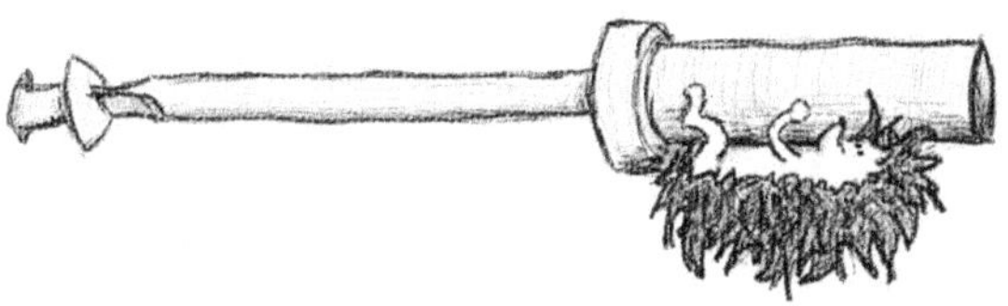

Dear Bunsen,

I've had a lightning stroke of realization. I have to prioritize my life. Weeks are passing, and I have made next-to-zero progress on the science fair, just some basic background reading on a couple topics, aerodynamics and telegraphs and crystals. I haven't built a thing. I've just been so swamped by everything else, planning and executing (the plan, not people). I'll miss my chance at the fair if I don't choose now and get cracking. But I can't neglect my villainy. Plus I still have that paper and regular homework ... things are piling up. So I'm going to proceed with the paper airplane project. Not flashy, but original and straightforward. Shouldn't take too much of my time, and should be enough for an easy win.

With the villain project on the side I figure best to go for the "clean and simple" approach. I'll make enough of a splash there. Focus on villainy: that's the priority.

Dear Bunsen,

No! Edith's gone and squirreled her way into helping hunt me down. Well, not officially me, the person behind the lunch volcano. Which happens to be me. She's going to start meeting with Mr. Murphy and Mr. Science, and tell them her observations as a student to help them. Give them new insights or something. Should have seen this coming, I guess, since she helps with everything else at school. Anytime, anyone, anything. Helpful, helpful, helpful! You'd think she'd run out of time. Or get tired. Nope. Not Edith. Always looking for ways to get involved. She's a monster! A popular, pretty, peace-loving monster. Who has now offered her services to my heroic duo. How obnoxious! Now I'll have to start calling it a "crew" of heroes. Or a "troupe" of heroes. What's the term for a bunch of heroes, anyway? "Team" is kind of boring. There's herds of cows, flocks of geese, murders of crows. A "protection" of heroes? A "revenge" of heroes? A "shield" of heroes?

Later:

You know, maybe I overreacted earlier. Now that it's no longer a shock I can see that Edith working with the teachers may even be helpful. She's bound to tell me all the news, which would put me one step ahead of them.

Dear Bunsen,

Another card from Alec! And he's heard about my work!

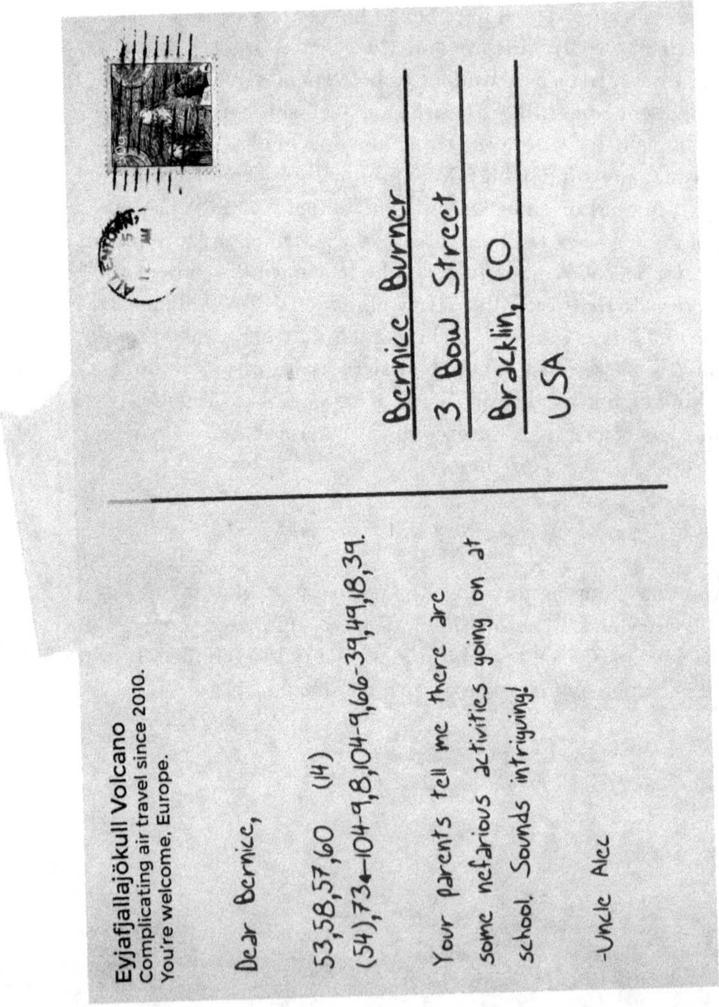

Eyjafjallajökull Volcano
Complicating air travel since 2010.
You're welcome, Europe.
Dear Bernice,
53,58,57,60 (14)
(54),73←104-9,8,104-9,66-39,49,18,39.
Your parents tell me there are some nefarious activities going on at school. Sounds intriguing!
-Uncle Alec
Bernice Burner
3 Bow Street
Bracklin, CO
USA

Dear Bunsen,

Laundry day is the worst. It just takes so much more time than any other chore. Sorting, loading the washer, switching it to the dryer, folding it, hanging it up, putting it away. And how, I ask you, is anyone expected to know how to fold a fitted sheet? The whole process is an all-day commitment, and meanwhile some of us have real actual work to do.

I did manage to get some outlining done for my English paper. I've come up with an idea that I want to run by Mrs. Reid — writing about the current events at school. I can look at the situation and analyze the heroes of the school who have risen to the challenge. Analyze and narrow down to who the hero really is. It's perfect! Isn't that a great idea? I just finished writing it all down in time. Edith started banging on the door, asking me to come with her to pick up some things from the store. Apparently we were out of mayo and ketchup. And baking soda of all things. Who runs out of baking soda? Unless it's the ketchup volcano culprit.

Villain Notes: Round Three

After some calculation, I've determined I need one more villainous teaser, to keep me in people's minds. Then I'll focus on how to declare myself and identify my heroic counterpart. By the end of the week I should come up with a good plan.

Dear Bunsen,

Do you know what gives eggs that delicious repulsive rotten-egg smell? Hydrogen-sulfide fumes. I looked it up today, because guess what? The kitchen smells like a hot spring.

It was my job to clean out the fridge today. At the very back, behind prehistoric leftovers and petrified condiments, was an ancient carton of eggs, dating from the days of the dinosaurs. I pulled them gingerly out, careful not to disturb any other ancient relics, but I may have held them a little too gingerly …

79-92 62, 4-5, $(3-53)^2$ He 74 79-92 19 4-5
90 $(66-39)^2$ ⇐ [4-5, 79-92]

Bunsen, never drop rotten eggs. I'm telling you now, it's a bad idea. What a putrid stench! The kitchen may never recover. I may never recover. My knees may never recover from scraping goo off the floor. My ears may never recover from the talking-to either.

And I learned an important lesson. Always empty the refrigerator from the front, not the back.

Villain Plan: Stink Bomb

Stink bombs are attention-grabbers. Anyone with a nose feels the effect. Unless they have no sense of smell. People with bad hay fever, for instance. Side note, I looked it up and it turns out there's actually a word for having no sense of smell: anosmia. Check it out:

"Anosmia: a partial or complete loss of smell due to blockage or damage. Duration varies by cause. Common causes include, but are not limited to: the common cold, influenza, acute sinusitis (sinus infection), nasal obstructions including bone deformity, and damage to or destruction of the olfactory system due to disease, injury, or aging."

Looks like it's not just people with stuffy noses. It can be permanent when there's damage to the "smelly" nerves from a traumatic injury or disease. Crazy!

Back to the point though, I can use smell to my advantage! I'll place a set of homemade stink bombs in the vents. Several, on a timer, to maximize the effect. I'll need matches, ammonia, and airtight bottles … I should be able to get what I need from Mr. Murphy's supplies. This one will take a couple days, so I have to move quickly.

Dear Bunsen,

The timer's proving tricky to plan. Something simple, but clever. I can't just go around shopping for timers. That's traceable! Someone could easily track me down if I just go around buying things willy-nilly. What I want is something like a magnet that would turn off after a given time, that I could make myself, but none of my books had a good answer. Time to email Uncle Alec again.

Dear Bunsen,
He came through!

From: Alec Burner
Sent: Today at 9:14 AM
To: Bernice Burner
Subject: DIY Timers

Bernice -

Sounds like you're looking for a homemade electromagnet. They're pretty easy, just a nail, a battery, and some wire, you can look it up anywhere. It's easy to calculate the amount of time it would last by dividing the life of the battery (amp hours) by current (amps), which you can calculate using Ohm's Law (current = voltage of the battery / resistance of the wire).

I'd start with a pretty long wire coiled up, and be careful to use gloves if you need to touch the wire during the first experiment, so you don't get burnt from a hot shorted wire. Wear some glasses too, in case it vaporizes. A short wire can get red hot even from a battery. I hate to say it but I know this one from sad experience. :)

I'm curious what you're trying to do?

Alec

Dear Bunsen,

It's all set. In three days we should be in business. I have to let the bottles sit, so the reaction with be potent enough. Of course, the timer's my favorite part, a stroke of brilliance. I'll set that up last-minute and it will all be good to go! And I got the tools. He doesn't even lock his door, I got in and out no problem.

Dear Bunsen,

Ha! Success! There was an emergency evacuation from school today! Well, okay, not the whole school. Plus, we weren't TOTALLY evacuated. And I guess it wasn't an EMERGENCY. But a whole hall of classes had to be relocated. In an unplanned fashion. And that sure feels like "emergency evacuation" to me. The reason? The air was filled with noxious fumes! "Fumes" as in a smell that permeated the whole wing, and "noxious" as in smelly beyond all reason!

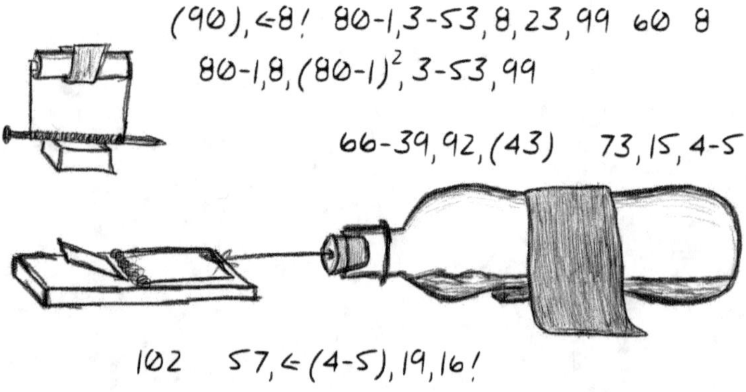

I was in one of the classes that had to be evacuated, Mrs. Reid's. She was giving a lecture, actually a pretty good one, about Hercules, Perseus, Theseus, Odysseus, and a bunch of other Greek heroes, and what the Greeks considered heroic, compared to what we might consider heroic. She was interrupted by sounds of gagging from the students near the door. Slowly and steadily, an invisible wave crept into the room, and row after row of students put their hands over their faces, trying to block the stench. I was one of the lucky few by the windows. Once it became clear that the smell wasn't going anywhere, we propped open the windows, despite the drizzle outside. Rather than dissipating, the smell stuck around, strong and persistent. Distracting, and disgusting. Rotten eggs. Yum. At first, Mrs. Reid tried to resume her lesson, ignoring the smell. But after a few minutes the air hadn't cleared, even with an open window, and when it became clear that the smell was not going away, she put her head out in the hall and found the smell even stronger out there. At that point, she had us gather our things and made a call to the office. I could only hear her side of the conversation, but I could gather what the office secretary was saying.

Mrs. R: "Hello, we have some kind of situation going in the west hallway. The air in here is foul, started a few minutes ago and is becoming worse, not better. I'm concerned that it is a danger to my students, and so I am taking them outside where the air is clear. Can we have someone down to check out the source? .... Yes, it's not just my classroom, it looks like the whole hall is affected, and the smell is worst in the hallways .... Okay, thank you."

"Okay class, let's head out. I know it's wet out, but better a little damp than in here, yes? They're going to have the other classes move out as well, and send someone to find out what's going on, try and get it so we can come back quickly."

As we walked out, all the other classes started filing out as well, everyone with hands over their faces, clutching their bags, hastily pulling them over their shoulders. It was chilly, but the rain had mostly stopped. I caught sight of Mr.

Murphy leaving the main office building with a box swinging at his side. The rest of class was a bust. Some of the teachers tried to resume teaching, but who can pay attention when they are cold and wet and outdoors? After a short while they moved all the classes into the empty cafeteria, and made an announcement that for the remainder of the afternoon all classes in the west hall would take place in the cafeteria.

Dear Bunsen,

Business as usual today. The west hall was usable again. But there was a buzz in the air. Everyone was speculating about the evacuation, what the cause was. No one knew anything about how they solved it though.

Dear Bunsen,

Mr. Science came through today; subtlety has never been one of his strong points. Lucky thing, too, because Edith hasn't been helpful at all. She's taken her new role as resident student spy very seriously, she has. Not a word about what the teachers are up to, not since she started meeting with them. But Mr. Science has likely never managed to keep a secret in his life. He's too excitable. He didn't say anything outright, but today's club meeting featured the latest news. He bounced in the door with his arms full of capped test tubes.

"Good afternoon, kidlets. Today we are talking about noses. I know you're all wondering, why do noses run and feet smell? We will unfortunately not be answering that age-old question today. But we will discover the power of molecular travel through the air! And that, my brilliant young prodigies, is a powerful knowledge to possess."

He talked us through the basics of smell, and then uncapped test tubes, one at a time, and had us identify them. Cinnamon, vanilla, banana, lemon, and then the last one.

As we wrinkled our noses, squirmed, and held our breaths, he grinned. "Breathe it in! Do you know what this is? That delicious rotten-egg smell is a concentration of hydrogen sulfide. Sulfur smell. A delightful element, sulfur. You can find this smell near hot springs, rotten eggs, and of course, most stink bombs."

Looks like they figured out the cause.

Villain Notes: The Next Step

Time to take things back a notch. Hold back until the fair, then blow it all out of the water. Two big statements is enough to grab some attention, and still stay under the radar. The last thing I want to do is to be found out because of a stupid mistake from trying to do too much. So it's time to pull back a little, make people think I've been scared away or that I've made my point and finished. Then come back with a flash and a bang! After all, according to *Supervillains 101: Protips to Becoming a Bane to Society*:

"The best villains are unpredictable. They spring out of nowhere, without predictability or pattern. Ideally, the mode of attack is also variable. It is understood that each villain should possess some distinctive traits, to distinguish him among his competitors. However, predictability leads to discovery and capture. Be aware of that when plotting."

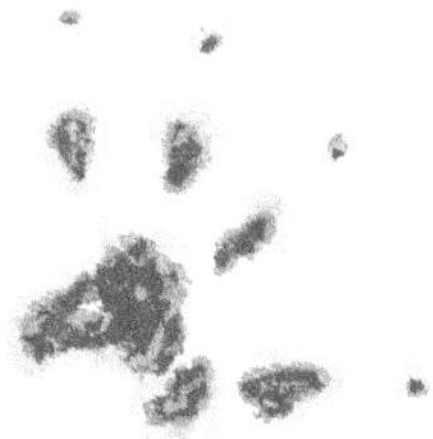

Dear Bunsen,

There hasn't been any kind of activity at school for the last few days, and rumors abound! Everyone's sure that the same person left the stink bomb and the ketchup volcano, and they're all waiting to see what happens next. But nothing has.

On a related note, they finally replaced Ms. Chu, with some quiet skinny lady — I haven't met her yet. She stays in the back of the kitchen mostly. But I heard Ms. Chu moved. Across the country. What an overreaction. I love it!

Dear Bunsen,

With how quiet it's been, I've had a ton of time to work on my science fair project. Finally! I've still got a lot of work to do, the report and all, but I've begun designing and testing, and it's all developing swimmingly.

I've almost wrapped up my heroes report, as well. I just have to get a couple more book sources to flesh it out. I was still able to use my original disastrous interview too. Turns out it wasn't a disaster after all. I'm using it as an anti-example, a way of showing how some people think they're heroes, and some people actually are heroes. It's really coming together nicely.

That's been taking up most of my time the past week or so. Not a ton going on in science club, just reviews to keep us on track for the fair, so I skipped this week since I don't have a ton of work to show yet. And I haven't seen Mr. Murphy in a while either. It's been pretty boring at school.

The coolest thing today? More mail from Uncle Alec! Another postcard, this time for both of us. How cool is that!

Villain Notes: Debut Plans

The best place for my big debut is obviously the science fair. Crowded, convenient, well-publicized: it's perfect! I won't reveal myself openly. Too risky this early on in my career. But I'll make a bang. A clear statement from a clear villain.

Snæfellsjökull Volcano

Fire & Ice- the erupting volcano clashes beautifully with the freezing surroundings

Image courtesy Otto Lidenbrock

Bernice and Bunsen-

13,42,16,22-53

26,(3-53)² 49 90

(116)←8,20,102!

How goes that project

of yours?

-Alec

Bernice Burner

3 Bow Street

Bracklin, CO

USA

Dear Bunsen,

When I got home today, Mom wanted to talk. I'd poked my head into her workroom to let her know I was in, and was halfway up the stairs when I heard her calling me back. She had that little crease between her eyes, the one people get when they don't realize they're frowning.

"Bernice, honey, I noticed you've been spending an awful lot of time cooped up in your room lately." She's worried I'm lonely!

"It's my lab, Mom. And I'm not cooped up, I'm very busy working on scientific developments. And my English paper."

"I'm just concerned, sweetheart. Is everything going okay at school?"

"Yes, everything's just peachy."

She clearly didn't believe me, because she launched into an interrogation session. I dodged what questions I could and hedged on the others, until she seemed satisfied. The last thing I want right now is Mom breathing down my neck. I'd never get anything done!

Dear Bunsen,

It's time for some recognition. Having discovered the best candidates for my nemesis, they need to start realizing their potential. Heroes need time in the spotlight, after all. The plan is to have Edith proofread my paper before I turn it in. With a skilled comment or two, I can easily persuade her that the heroes deserve some recognition for their heroic deeds. Easy as pie. She'll waste no time bringing it through the proper channels and making that happen.

Villain Notes: Update

Things have been going well so far. My plans have been working without a hitch, though I'll admit, the last trick was a challenge to pull off with Mr. Murphy and company at the helm, road-blocking me.

It's time for Phase Two.

Everyone's seen a taste of what I'm capable of now, and the school's on red alert, with a clear team of suitable heroes to take me on. The perfect setup for my debut. After that the world will see the hero it needs when I take the stage in Phase Three. Easy peasy.

Dear Bunsen,

It's nearly midnight, but Edith just got home from her last away game of the season. I was putting the final touch on my paper and heard her plod up the stairs and thump her duffel on her bed. My door squeaked and I heard a whisper.

"Bernie?"

"Yeah, Ede?"

"I saw your light, and wanted to say good night. Working on homework?"

"Yeah, just wrapping up my paper finally."

"Sorry you have to be up so late. I can edit it tomorrow if you'd like. Love you."

"Love you too. Good night."

"Night." She inched the door shut.

Bunsen, I don't know what possessed me, but I heard myself whisper, "Edith?"

The crack widened. "Yeah?"

"Did you win tonight?"

"Yeah. We won."

"Good. Sleep well."

"You too."

That girl is just too nice for her own good. At least I didn't have to ask her to edit my paper.

My Conversation With Edith About My Paper
"Bernie?"

"Yeah?"

"You put me in your paper. You really think I'm one of the heroes of the school?"

"Well sure. I mean, look at how you're helping Mr. Murphy and Mr. Science. That's more than anyone else is doing."

"Wow, Bern. That's the nicest thing anyone's ever told me."

Dear Bunsen,

An announcement at school!

"Attention students. We will be having a special assembly tomorrow immediately following lunch. Afternoon classes will be cancelled. Report directly to the gymnasium after the lunch hour. That is all."

Looks like Edith pulled it off!

Dear Bunsen,

Read over my English paper again tonight and realized something. There's an inherent logical flaw in my approach that I've overlooked. Or forgotten. Or something. Not in the paper—everything I said in there is true. But I've discovered something about my villainous situation.

In the paper I naturally discuss three different heroic people: Mr. Science, Mr. Murphy, and Edith, who I had to include because she's helping, even though she's not as cool as the other two. I've been using that to guide my actions, taking the heroes and making situations where they'll rise to the occasion. Like any villain would do, right?

Time for a reality check. Bunsen, I can't deal with three people at once. Not long-term. That's too many. It doesn't make sense that one girl can defeat two adults and her older sister. Be realistic! I'm good, but no one's that good! And on that note, a kid against even one adult is still pretty ridiculous. Even if Mr. Murphy is the perfect hero. So I'm going to have to reassess my situation and come up with a better plan. At least until I'm a bit further along in my villainous career and can handle a team.

That being said, I still think the assembly is a good idea. May as well give credit where credit is due.

Dear Bunsen,

The assembly was today. Here's how it went:

The gym slowly filled up with students. My class got there early, so we sat a couple rows back, pretty near the front. Plenty close enough to see everything that went on. Edith was dressed up, sitting on the stage by Mr. Murphy, talking animatedly to him about speeches while she went through her note cards. Mr. Murphy seemed even more quiet than usual, hardly responding at all to Edith's chatter. He seemed a little nervous, sweat beading his brow as he looked at his hands in his lap. Probably thinking over what he would say. The principal strutted in after a few minutes and sat down in the empty chair by Edith, greeting both. Slowly the room filled, and with it grew the roar of dozens of whispered conversations, drowning out the conversation on the stage. After a few minutes, the principal stood and greeted us.

84,8,104-9 (95),7

"I'd like to welcome you all to this special assembly today. As you know, we've been monitoring the 'science bandit' situation closely, and are pleased to announce that thanks to some real ingenuity and effort on the part of the staff and students, the vandalism has died down. This is in good part due to the effort of our head of maintenance and janitorial, Mr. Dan Murphy. We are pleased to commend Mr. Murphy and the student council today for their combined work and diligence. Without further ado, I present to you Dan Murphy and Edith Burner, who is representing our student council. Dan, Edith?"

Applause. Edith smiled and stood, the picture of ease. Mr. Murphy took a deep breath, glanced toward the door, and stood beside Edith. They walked to the mic together, Mr. Murphy's face getting pinker and pinker.

"Dan, on behalf of the school, I'd like to personally thank you for your efforts these past weeks. You've done the school a great service." The principal offered his hand, and Mr. Murphy shook it uneasily, then started backing towards his seat.
"Before you go, Dan, is there anything you'd like to say?"

Mr. Murphy cleared this throat clumsily and walked back up to the mic. "Um, thanks, everyone. I just try to do my job, that's all." He stepped away, and there was a smattering of confused applause as he sat back down, fingers clenched together.

The principal's next words faded into the background. Mr. Murphy sat staring at the floor, out of his element, refusing to acknowledge his contribution. As Edith graciously shook the principal's hand and stepped up to the mic, I realized something, Bunsen. Something important. Superheroes have to live a public life, protecting the universe from the villain. That requires stage presence. It comes with the job. Accepting awards and smiling after a battle. Heroes can't have stage fright! Mr. Murphy doesn't do well in the limelight. What a good thing I changed my mind about him being my hero.

Award Ceremony to Honor School Heroes

CAFETERIA— This week, the school celebrated the hard work of Mr. Dan Murphy, Mr. Steve "Science," and basketball star Edith Burner, who each aided in the latest school effort against mischief. These three led the charge, hunting down the mysterious prankster who has put everyone through the ringer with his dastardly deeds. His disappearance is attributed to their heroic efforts.

Dear Bunsen,

I was re-reading *Nemeses: The Origin of a Pair*, trying to figure out how to fix this mess I've gotten myself into, when I stumbled upon inspiration. An epiphany! The solution is so simple, so obvious. The book pointed out how often nemeses will have a backstory, some common element from their past that brings them together later. They grew up together, that kind of thing.

That's when it hit me like an apple to the head. I do have the perfect candidate for a nemesis. And you're not going to like it.

Yep. Edith.

Think about it. She's already gotten into the hero gig, and has mentors to help her grow into it even more, Mr. Murphy and Mr. Science, and even Mrs. Reid, the resident expert on all things heroic. She's young and inexperienced like me, so she can't outwit me easily, and people already know and like her, so she won't have to win people over. Mr. Murphy and Mr. Science will naturally still be involved, but I can change my tactics to target Edith, so they become her supporting cast. Like policemen and firefighters support superheroes.

Trying to keep it a secret from her will be hard, you know, since we live in the same house and everything. That's a concern. But I can handle it. I've done well so far, haven't I?

A manageable, moldable hero. With me as the puppet master behind it all. I think I finally figured it out.

36

Dear Bunsen,

I was curious if Edith's still worried about the events at school, now that they've died down some. She hasn't said much lately. Actually, I haven't seen her much lately. I've been pretty focused on getting my science project together finally. Just need to finish the poster and I'm set.

Anyway, the conversation went something like this:
"Man, these pranks are getting out of hand. They seem sort of sinister, maybe even villainous, almost the work of a criminal mastermind, don't you think, Edith?"

"Well, I wouldn't say sinister or villainous. They're annoying, and they're a threat to authority, which is definitely not good. But I don't know if I'd say villainous is the right description. Plus, it's died down."

Just you wait, Edith. Just you wait. You'll see.

Dear Bunsen,

I was talking to Edith while we put groceries away for Mom today. The conversation meandered its way to discussing our projects … I need to know how I can set her up as a hero, and if I can use her project, that will be perfect.

Edith: So I know you had a few things in mind for your project, what did you end up deciding? I haven't seen you work on it much yet.

Me: I've been working on it, but it's not as involved this year, very simple to set up and test. I'm investigating the effects of different designs of paper airplanes on their velocity and flight duration in a wind tunnel. The tunnel of course gives me a controlled environment in which to compare the planes.

Edith: Oh, really? That's kind of toned down for you. Last year you went all-out with the disappearing ink. I kind of expected you to do something along those lines again, since it was such a good idea.

Oh, if only she knew.

Me: Yeah, well it's been pretty busy, so I figure do a clean smaller project and do better than a half-baked larger one, right?

Edith: Right, that's very sensible of you.

Me: So what did you decide to do?

Edith: Oh, well, I decided to make a lie detector.

WHAAAAT?? Who thinks of that kind of thing??

Me: Oh, a lie detector. What inspired that?

Edith: Well with all the weird things going on at school, I was thinking about people sneaking around, and lying, and then I ran across some instructions on how to make a homemade lie detector. It's really cool, apparently when you lie your fingers sweat, so it measures the difference in wetness ...

She continued, but I was too busy laughing inside to listen. It sounds interesting, but I'll have to look it up later. It's just too ironic. My nemesis, making a lie detector for her science fair project. Ha!

I have to admit, my project is going to look pretty shabby next to that. I mean, airplanes are cool, but a LIE DETECTOR? I'm wondering if I should try to build up my project a bit. But how?

Dear Bunsen,

What do I do, what do I do? I don't think my project has a chance of winning against Edith's. If mine were a bottle rocket, hers would be a space shuttle. No matter what I do, it's too late to turn a bottle rocket into a space shuttle. I've been racking my brains, but prodigious as they are I just can't see a way to win. Sabotage is unethical. Plus, then everyone would feel bad for her, and I want her to be admired, not pitied.

Hey wait ... I'm having an epiphany. I want her to be admired ... maybe ... maybe that's it. I've been so focused on wanting to win that I missed the big picture here. What should be my number-one priority? A school science fair, or my future as an awesome villain with a perfect nemesis? And what, I ask, could be more perfect than the winner of the science fair rising up to challenge a science villain?

It's brilliant! I can't believe I didn't see this before! If only I'd realized earlier. I wasted so much time and energy trying to do my own project along with everything else. It's probably a good thing long-term, it gave me a good cover of normalcy. People would have gotten suspicious if I stopped doing the fair. But man, good thing Edith is doing the fair after all — what a lucky break, it's beautiful! She'll win, everyone will see how perfect she is for the job, my victory will be complete! Wow, I can't believe I almost ruined Edith's heroic chances by defeating her in the science fair. What was I thinking??? I wonder if she needs any help finishing up. I want her project to shine, after all. Maybe I'll go check in a minute.

But first, one thing I still need to manage is how to sneak in and prepare the scene for my big moment. I still need to perfect the climax: find the best way to inspire shock and awe, and work out how to ensure Edith will be the one who steps up. She'll want to help of course, but this has to be airtight. No mess-ups.

Science club is in charge of setting up for the fair. I can manipulate my time there, I'm sure. Once I work out the final details, all that's left is to wait and watch until the perfect moment arrives. Then boom!

<u>Possible Hidden Messages:</u>

"This is only the beginning" - of what?
"Prepare for disaster" - lame.
"Sciences rules - and so do I!" - better...
"Science rules the world. I rule science. Bow before me!" - too dramatic, yeesh.
"No one is safe."

Dear Bunsen,

Just when I sat down to brainstorm the perfect climax for my debut, I checked my email, and there was my answer! I didn't even have to ask for anything this time.

From: Alec Burner
Sent: Today at 3:56 PM
To: Bernice Burner
Subject: (no subject)

Bernice - Look up NI_3. You won't regret it.

- Alec

Dear Bunsen,

I've been thinking and planning and scheming and planning and conniving and planning, and I couldn't figure out the best way to pull off my plan without getting nailed. But I think I got it.

Here's the plan: Science club sets up for the fair, at which point the room will be filled with tables, and the "Welcome to the Fair" banner will be hung. We painted that one at last club meeting. Then everyone will come and set up their projects, and Mr. Science will walk through and check that all the rules are being followed. That happens tomorrow night. The actual fair is the next day, all day long.

I have "borrowed" some tools from Mr. Murphy's vast supply, including a drill, screws, and rope, which I will use to secure the successful bucket drop. Note to self: owning tools will come in handy one of these days. I can come back after the tables are set up and do it then because no one will expect anyone to come before Mr. Science has done his runthrough, and once it's in place, no one will see it. I can also make my additions to the banner at that point.

The next bit will be trickier. I'll have to come back early morning before the fair for that part. The trick will be placement. I'll have to check that the angles will be right, but it should work to paint right by the banner, and it should be protected from being stepped on by the tables underneath the banner. My only fear is that it might go off before I'm ready. It shouldn't make a HUGE difference, but I want the timing to be right. More dramatic that way, you know? And if I'm going to make a bang, I want to make it right!

Dear Bunsen,

A successful day. Somehow I managed to rig the bucket, paint the banner, position my project, and smear the explosive with no hiccups. The dropcloths on the tables hid the black sludge from view, the ink on the banner dries clear, and no one ever looks at ceiling tiles, so once everything was in position, there was nothing suspicious to see. The timing was a little tricky, I'll admit. I had to keep running in and out, to divert attention from myself. After all, it doesn't do to get caught by acting suspicious. So I would go in and do one thing at a time, while watching for other kids coming in to set up. Then I'd leave for a bit, let them be alone in there, while I went and fetched the trophies, or printed the judging sheets. Perfectly innocent activities.

The banner was easy — someone else painted it, and all I had to do was volunteer to bring it in. Preparing it took no time at all.

Being on fair setup meant I could decide where projects went, so positioning my project was a cinch.

It took some finagling to rig the bucket. At first I wasn't quite sure how to accomplish that — I'd need access to a ladder, and how on earth could I bring in one of those inconspicuously? "Uh, yeah, I just need to … put a bucket in the ceiling … don't worry about it." But then I realized, the banner! We could easily hang it from the ceiling. Boom. Ladder, check. Reason for being on the ladder, check.

The last bit, the big bang surprise, a few of us had to pop under the tables a few times to tape things down and whatnot, and painting in the corners hid the black sludge just fine. The power of camouflage.

So it all came together quite nicely. The fair was put together, my debut was put together, with no one the wiser.

Dear Bunsen,

It's a miracle. Everything went off without a hitch.

The day dawned bright and beautiful, perfect for some evil. Edith and I arrived a little early to check in, and the cafeteria was soon packed with all the other students dressed up all professional waiting for their turn with the judges. That part was slow, just waiting around. But a little later the parents came for the public viewing while the judges tallied up the marks. I struggled to concentrate on people's questions, even during judging, which I'm sure didn't improve my ranking. I was too distracted. But eventually the moment came! And that was the best part.

Right as Mr. Science stepped up to the mic to announce the winners, the ceiling opened, dumping a pail of grape juice and a rain of marbles onto the banner and on the floor. The marbles bounced and rolled under the tables, where there was a flash and bang as purple smoke billowed. The banner started changing color as people began yelling, and the message showed up nicely. I went with "No one is safe" and a picture of an Erlenmeyer flask. Classy, don't you think?

People started panicking, and then I heard Edith grab my wind tunnel off the table, take the fans off the ends, and direct them to blow the smoke out the window. Mr. Science tried to maintain some control over the situation by directing people outdoors, but despite his efforts, chaos reigned as people scrambled for the exits. Because of me.

What a glorious day.

SCIENCE FAIR
WELCOME

NO ONE
IS SAFE

Dear Bunsen,

At this time I feel it is appropriate to briefly reflect upon my journey to villainy these past months. Such musings naturally require a tone of formality and dignity. I shall oblige. Truly, one might find it remarkable that a girl not old enough to drive could accomplish so much in such a brief period. It is shocking to consider that in a few short months a nobody could disrupt the lives of so many, driving people to panic and dismay, even causing some to depart for gentler climes.

...

Okay, I can't take myself seriously writing like that. Bunsen, isn't this cool? We've done so much! There have been hiccups since I'm learning as I go, but no fiascos, no disasters. Everything has sorted itself out for the best, even things I thought were horrific. Like Edith joining Mr. Murphy and Mr. Science. If she hadn't done that I wouldn't have had such a perfect hero at my disposal. It's amazing!

Still waiting to hear the official view from the school about my success. They locked down the cafeteria for the rest of the day to investigate, and that's really all the news so far. Mom and all the other parents received an email from the principal explaining that there was an act of vandalism which led to an evacuation, but there were no injuries and an investigation has begun. In the meantime, the whole cafeteria is sitting as-is. The banner, the marbles, the posters and experiments. Nothing moves until at least Monday. A shrine to my brilliance.

Dear Bunsen,

We had permission to remove our experiments this morning, and when I returned to pick mine up I overheard some delightful conversations. All the other kids were talking about the explosion and what happened on Saturday:

"... he just got up and then, BOOM! I didn't scream, but this kid next to me ..."
"... and my clothes are all stained pink from the splash ..."
"... I was this close to getting hit by a marble ..."
"... did you see when the sign changed?"
"... look at it now ..."
"... all that purple smoke, I thought I'd die ..."
"... you can still see where that stuff exploded ..."
"... and did you see Edith? She was so fast, no hesitation ..."
"... I bet they give Edith a medal or something ..."

And then this little gem:
"Do they have any idea who it was?"
"I bet it's that same kid from before. But didn't they catch him weeks ago?"
"No, stuff just stopped happening so they thought he gave up ..."

Keeping a straight face in there was one of the hardest things I've ever had to do.

Dear Bunsen,

Finally, finally, the news is out and it is beautiful. First off, I got my paper back from Mrs. Reid. At the end she had written this in purple ink: "Nice job, Bernice. A great paper on some very local heroes. You put a lot of work into this, and it's clear to me you learned a lot in the process of planning and writing. Well done!"

More importantly, they finally announced the science fair results this morning: "Due to the events at the science fair, we were unable to announce the winning results on location. However, we are now pleased to announce that the first-place award goes to Edith Burner, with her homemade lie detector. Second place goes to …"

Yes yes yes! All according to plan!

But most importantly, the school paper came out, making it all official.

"Cracked Chemist" Back with a Bang at School Fair

GYMNASIUM— The science fair is a time of innovation and investigation, excitement and nerves, where students eagerly wait to show their projects and receive the scores that will determine whether they win or lose.

It all went up in smoke this weekend with the reappearance of the elusive prankster who is now being called the "Cracked Chemist."

The fair dawned as expected. No sign of any mischief. But announcement of the science fair winner was interrupted by a rain of purple juice which painted the wall, floor, and banner, followed by a series of explosions, which our very own Mr. Science has identified as the reaction caused by a compound known as nitrogen triiodide, which explodes on contact, and left everyone choking, disoriented, and dazed as they were engulfed in reddish-purple smoke.

Witnesses tell of the heroic actions of Edith Burner, the well-known student body president and basketball star, who once again showed her ingenuity and speed by grabbing two fans from a nearby project, using them to clear the smoke enough for people to evacuate. "Edith was so quick," her sister Bernice said in an interview. "Everyone was just standing there kind of shocked, and she just swept in, grabbed my fans, and really saved the day. If these nefarious activities continue, we're going to need heroes like that, don't you think?"

Ms. Burner's actions were followed by an immediate evacuation of the fair.

One source recalls the event: "Well, so they had just called for everyone's attention to announce the winners, stood up on this little stage by the microphone, with their list of names, and then a crash, and splash, and then there's the floor and wall and banner all wet. And everyone is looking around, and I can see the banner is starting to look funny, but then bang! bang! purple smoke everywhere! Then I hear 'We would like everyone to make an orderly exit' and no one can really see, but we're all jostling and there's babies crying and it was crazy!"

The banner had a hidden message revealed by the juice, "No one is safe." Speculation is that the message was painted in baking soda the night before the fair, and the bucket rig was set to spill by means of a remote control.

[A] hidden message revealed by the juice, "No one is safe."

The results of the fair have been withheld temporarily, but will be announced sometime this week, and in the meantime, the school is on its guard for more actions by the Chemist. One can hope that Edith Burner and others will be able to rise to the occasion again should the need present itself.

We did it. My search for a nemesis is complete. Let the games begin.

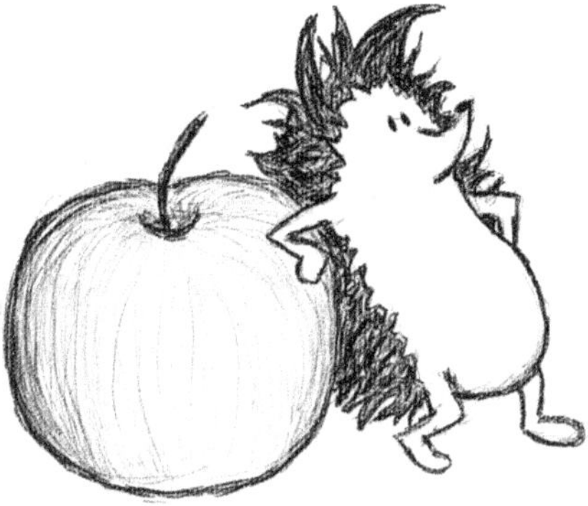

Dear Bunsen

Ask him a question at
www. ElementsOfEvil.com

More Bernice.
More Bunsen.
More Evil.

Read the continuation of Bernice's story in the second book of the Elements of Evil series!

About the Author

Brooke's fourth-grade science fair project was an attempt to bring together all the things she loved then: "Does reading improve your drawing skills?" The project was inconclusive, but it did teach her one thing: spilling vinegar on washable marker drawings can ruin your results. Thus began a long career of doodling, reading, and science-ing, which finally led her to write about Bernice and Bunsen. Her daytime job as a software engineer and master's student keeps her busy, but she loves playing writer/illustrator between classes.

www.ingramcontent.com/pod-product-compliance
Lightning Source LLC
Chambersburg PA
CBHW071215070526
44584CB00019B/3032
9781634110051